CAPITOL REEF

THE STORY BEHIND THE SCENERY®

by Virgil J. and Helen Olson

The late **VIRGIL AND HELEN OLSON** collaborated on writing this book while Virgil was assigned to Capitol Reef. Virgil was a 24-year career employee of the National Park Service, and Helen was a Special Education teacher.

A bold river flowing through a towering
canyon, huge white domes
topped by a brilliant blue sky,

green beauty in a vast desert seemingly
mysterious formations carved by
whimsical forces—all this is Capitol Reef!

The Capitol Reef Story

JEFF GNASS

The name Capitol Reef combines two distant and seemingly unrelated subjects: "capitol" known for the white domes of Navajo Sandstone that resemble capitol rotundas along the Fremont River, and "reef" for the rocky cliffs, which are a barrier to travel, like a coral reef.

This long narrow park is a geologic masterpiece. The up-lift, bending, and folding of the many layers of earth yield evidence that rocks can be tortured, by the same forces that created them.

The prehistoric Fremont Indians, as they are now called, inhabited the Capitol Reef area from approximately A.D. 700 to A.D. 1300, a period of 600 years.

Why did they come? The answer is simple—water. The lush, green vegetation of the Fremont River Canyon contrasts sharply with the desert to the east and the sagebrush plains to the west. Water is the key element to virtually all the parks in the great southwest. Water is what carved the earth to make Capitol Reef look the way we see it today.
Water determined what kind of plants and animals we see in the park.

Water meant food and life to the Fremont Indians, when it diminished, people moved on. Even today, rainfall is changing. Everything, everybody adapts.

Erosion is an ongoing process. Melting snow in *spring and summer rains wash minute particles into creeks and rivers to be carried to the Colorado River. Here the silt-laden Fremont River takes on a milky color during the spring runoff.*

GARY LADD

Hiking is the best way to get *really close to the formations and the plant/animal habitats of Capitol Reef. Much can be experienced even on short hikes. Park rangers can help you plan a hike that will suit your time, interest, and ability.*

Visitors today see all the forces of nature at work and in harmony. Birds and animals traverse the land in keeping with the availability of water, and the things that grow throughout the four seasons.

Capitol Reef does have four distinct seasons; the snows of winter add a new dimension to the colorful bands of rock. Spring—summer and fall are the sequence that begets the flow of life in all its form.

Even an orchard of historic value is maintained by the park service as a way of showing how the pioneers used all the aspects of nature to provide their way of life in this remote land.

As you come to see all phases of the park, take the time to see how they all fit together in balance—at Capitol Reef National Park.

the Term REEF is BORROWED

The resemblance of the rounded, grayish-white domes to those of capitol buildings inspired the "Capitol" in Capitol Reef. "Reef" was simply a nautical term carried over into geology to describe a barrier.

Eons Past

GARY LADD

Lapping waters of a tidal flat left *these ripple marks millions of years ago. Then, tilting of the earth's crust caused a shallow sea to shift in a northerly direction. Subsequent uplifting, fracturing, and erosion of the area has left a scene of almost unbelievable chaos—a challenge for geologists and a treasure for sightseers.*

Seventy miles long, this area of sheer multicolored walls and deeply eroded canyons encompasses nearly the full length of the Waterpocket Fold, and covers a span of some 250 million years.

The geological story of Capitol Reef involves many complex factors. Water is probably the one contributing most to the processes of deposition and erosion. Ancient seas cemented the limestones, shales, and sandstones as deposition occurred. Rivers carrying tremendous silt loads ground away at rocks and cliffs, carving canyons and valleys.

Ice created pressures which forced rocks apart, and glacial ice pulverized rocks and boulders, making them part of the soil. Raindrops softened and eroded the hard materials, and transported them to other areas. Water, in its many forms, was the tool which chiseled and sculptured the landscape, leaving it as it is today.

JOHN P. GEORGE

The prominent *escarpment that forms the western edge of the Waterpocket Fold along the road to Capitol Gorge is stunning. The graceful, green orchards of Mormon pioneers nestle in a red rock chaos.*

***It may take thousands of years for** the "Twin Rocks" to topple off their pedestals, but due to the unrelenting erosional processes of nature, we know it will happen. The boulder in the foreground appears to have shared the same history as the two still standing. The pedestal finally wore too thin to support its tremendous weight.*

JEFF GNASS

THE FORMATIONS

The exposed formations of Capitol Reef start where the north rim of the Grand Canyon ends. The oldest rocks exposed were there before the beginnings of life. There is actually a slight overlap, because the top two strata of the Grand Canyon—the White Rim Sandstone and the Kaibab Limestone deposited during Permian time (about 250 million years ago)—are also found here at Capitol Reef.

These formations were laid down when a large shallow sea covered the Capitol Reef area, as it did many other sections of the Colorado Plateau. Then, due to downward tilting of the land toward the north during early Triassic time, the Capitol Reef area became a broad, flat flood plain, then a tidal flat, and back again to a shallow sea.

Consequently, the Moenkopi Formation with its chocolate-colored sandstones, siltstones, and mudstones was deposited. Ripple marks and mud cracks indicate floodplain and tidal-flat deposition.

Reptiles and amphibians were roaming the country at this time, swimming in shallow waters and leaving their prints in the sandy bottom. As one looks up beneath the ledges in the Moenkopi, the casts of these prints can be seen in many places.

Although the Shinarump is a member of the Chinle Formation, it is significant enough to describe separately. This yellowish sandstone was deposited in stream channels, and is not found in continuous

Reptiles and Amphibians were Roaming the country at this time

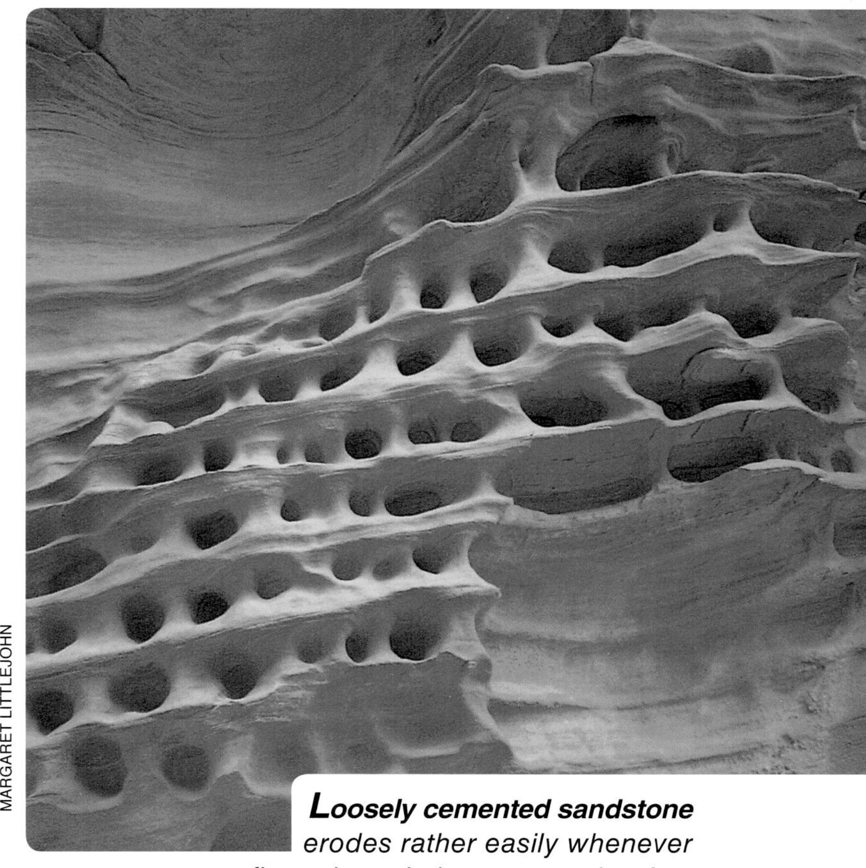

MARGARET LITTLEJOHN

Loosely cemented sandstone *erodes rather easily whenever water flows through the canyons. Ice Age meltwaters deepened canyons quickly.*

JOSEF MUENCH

For some, the Egyptian *Temple, at the base of the great escarpment, resembles the lines of a great temple in the Valley of the Nile. The soft Moenkopi has been protected by the harder Shinarump Sandstone on top, and has formed a cliff-like wall. The white substance in the walls of the Moenkopi is gypsum. Sparse desert vegetation consists of pinyon and juniper trees, and various species of saltbush and hardy grasses.*

TOM TILL

In the continuing saga of *nature, one can again see evidence of hard rock covering softer materials. Chimney Rock provides another of nature's whimsical examples as a small portion of Shinarump has remained to protect underlying Moenkopi. Once the remaining cap is gone, the rest of the "chimney" will erode rather rapidly (geologically speaking). From the Chimney Rock parking area, a foot trail leads up onto a small plateau where many exciting discoveries await the discerning eye, and exceptional views for photography abound. Tracks of extinct reptiles are found in the Moenkopi.*

layers. The Shinarump is a hard sandstone, and in many places it protectively covers the underlying soft Moenkopi, producing formations such as Egyptian Temple and Chimney Rock.

At the time the Shinarump and other members of the Chinle were laid down, the vegetation was lush; consequently it is now a place where petrified wood can be found, dull brown, tan, and sometimes black in color. Petrified Forest National Park is situated in the same strata, but there the mineralization is much more colorful.

Finally, it is the wood in the Shinarump that collected uranium as it percolated down through the layers above, and it was the Shinarump which was the main target for prospectors during the "uranium boom."

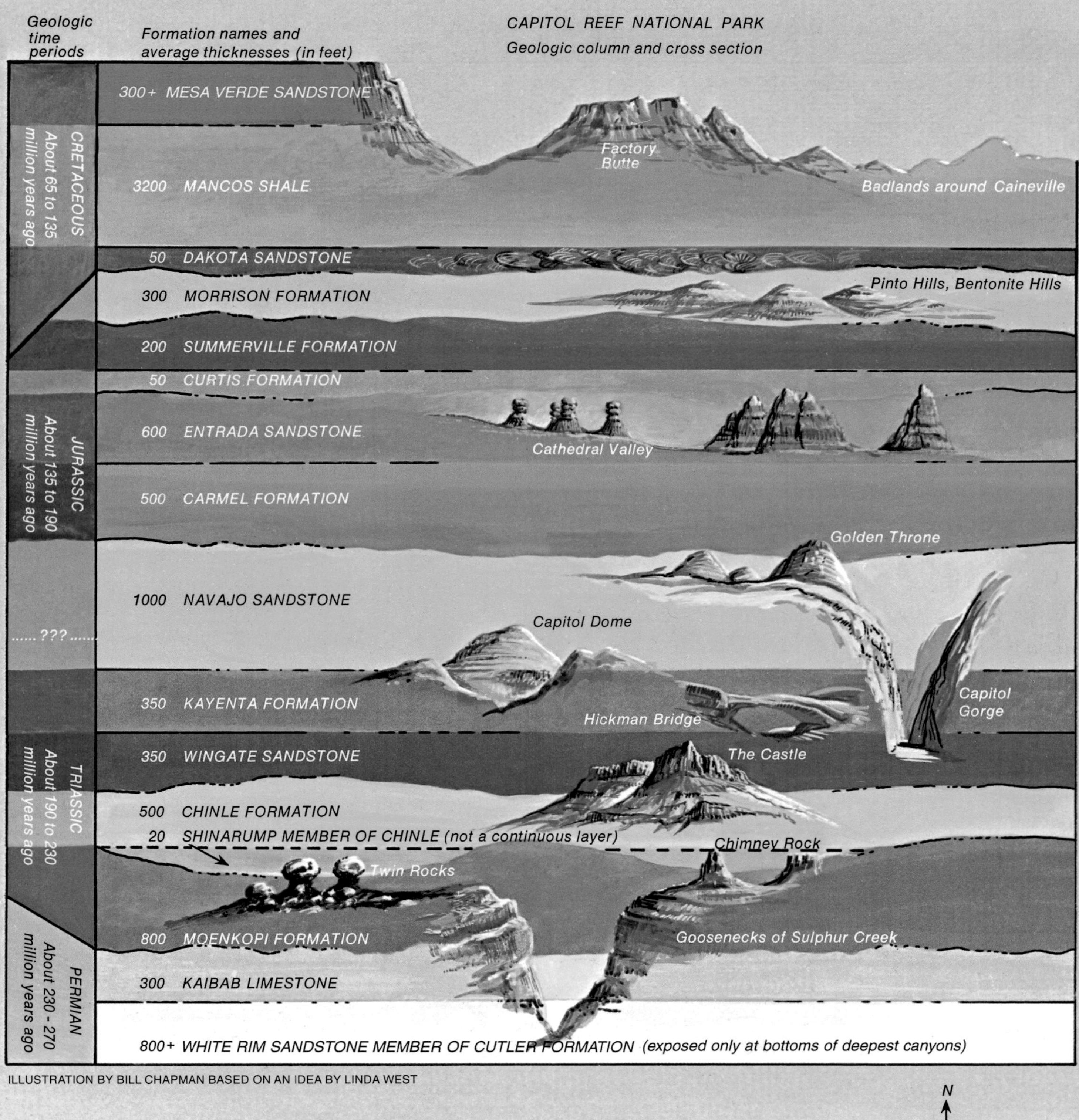

ILLUSTRATION BY BILL CHAPMAN BASED ON AN IDEA BY LINDA WEST

Irregular lines between formations indicate unconformities between strata; that is, breaks in the rock record represent periods of erosion rather than deposition. Named features shown above, such as Chimney Rock, are placed in their stratigraphic positions only, not in their actual geographic locations. Features in the cross section below, however, are shown approximately in their proper east-west relationships.

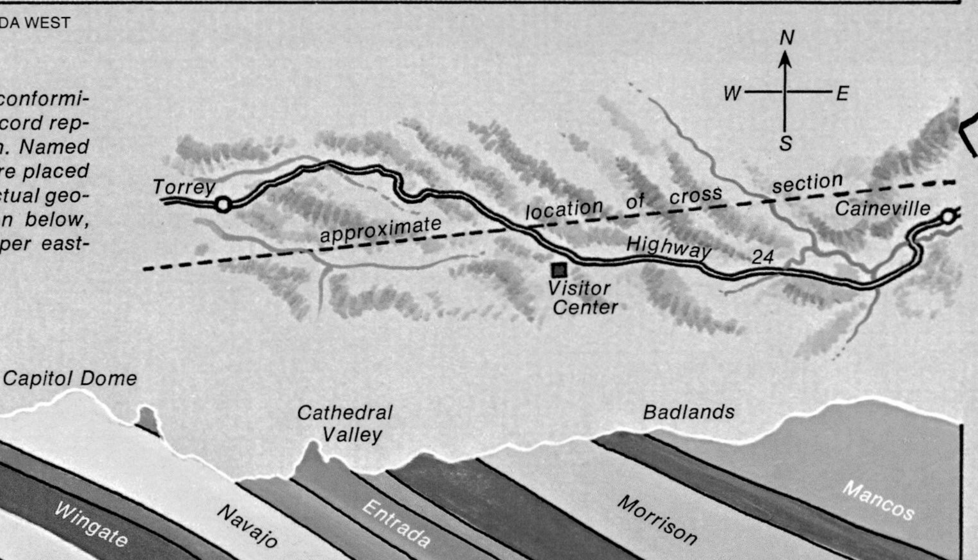

Colorful beds of the upper members of the Chinle—purples, greens, grays, and browns—are a combination of sandstones, silts, and bentonite (volcanic ash) laid down much the same as the Moenkopi—in broad flood plains and streambeds.

During the late Triassic, 190 million years ago, the land again underwent changes. Another uplift further drained the shallow sea, and the area became a desert in which blowing sand drifted into huge dunes covering large sections of land.

Three sandstone formations were laid down during this time, mostly by wind, some by streams. The first can be seen in the high, straight, reddish-brown cliffs of Wingate Sandstone. Erosion of the soft underlying Chinle causes the Wingate to fall away in large blocks, which can be seen scattered about on the slopes of the Chinle. (Many of these blocks lie across the highway from the visitor center.)

When the area was again tilted, stream action deposited the Kayenta Formation directly on top of the Wingate. These two formations are so nearly the same color that it is difficult to tell them apart.

LARRY BURTON

C**ottonwood trees enhanced *by fall colors provide an artistic frame for the colorful formations in the background. With the Moenkopi at the bottom and the Navajo Sandstone on top, can you pick out the formations in between as shown in the diagram on the preceding page?*

JOSEF MUENCH

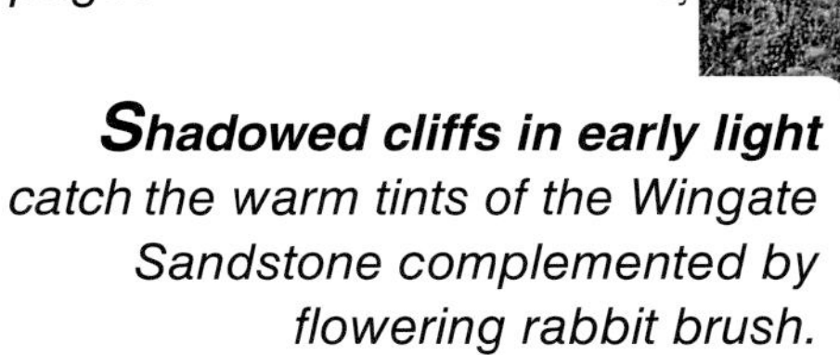

S**hadowed cliffs in early light *catch the warm tints of the Wingate Sandstone complemented by flowering rabbit brush.*

GARY LADD

***N**avajo, Kayenta, and Wingate sandstones* *are probably most responsible for the rugged scenery of the Waterpocket Fold. The Navajo for the domes and peaks that dominate the skyline; Kayenta and Wingate for the spectacular walls that enclose the canyons and build the towering cliffs. Very little vegetation grows in these formations except on flat surfaces where a little soil has accumulated, or secluded nooks where a little moisture seeps out of a hidden spring.*

***T**his river-deposited formation, the Carmel, was* *laid down on top of the Navajo Sandstone during the early Jurassic around 135 million years ago. The deep maroon color of Carmel contrasts strikingly with the light gray of the Navajo, and is most easily seen in the more southern area of the park.*

MICHAEL S. SALAMACHA

The last of the three sandstone formations is the massive, 800 to 1,000-foot-thick Navajo Sandstone, laid down, again, in a desert sand-dune environment.

The resemblance of the rounded, grayish-white domes of the Navajo Sandstone to those of capitol buildings inspired the "Capitol" in Capitol Reef. "Reef" was simply a nautical term carried over into geology to describe a barrier. The Waterpocket Fold, running nearly 100 miles from Thousand Lake Mountain to the Colorado River, is indeed a barrier; it can be crossed in only a few places.

Following the deposition of these three sandstones, the land again tilted—this time toward the south. During the next period, a shallow sea advanced and retreated many times. Rivers flowed from this sea into the lower areas. The resulting deposits of limestone, gypsum, sandstone, and claystone created the Carmel Formation. After this period of unsettled activity, the land was covered by a series of calm, land-locked basins in which the soft, reddish-brown Entrada Sandstone and the hard, gray Curtis Sandstone were deposited. The combination of the soft but massive Entrada below, protected by the harder Curtis above, has produced the impressive monoliths of Cathedral Valley.

A third sandstone formation—the Summerville, with its thin, even bedding much like a lighter version of the earlier Moenkopi—was deposited in this shallow-lake situation. Beds of gypsum in layers one or two inches thick, common in the Summerville, give it a striking appearance.

Following this period of quiet deposition, the story repeated itself and the land again lifted to create a wide flood plain which eventually settled to a swampy lake. During this time the colorful Morrison Formation with its reds, purples, greens, yellows, and grays was deposited. This formation was laid down in the late Jurassic, approximately 140 million years ago, the age when reptiles were most numerous. Fossil remains of the giants of the past have been found in these layers here at Capitol Reef.

LARRY VENSEL

Erosion takes many forms in the Navajo Sandstone, Domes, *pinnacles, and almost any shape the mind can conjure up. This formation stands atop the Fremont Canyon on the west side.*

The Entrada Sandstone cliffs of the upper Cathedral Valley *were very much a repeat of the situation in which the earlier Moenkopi was deposited—a time when the area had again tilted back to an almost level configuration with many land-locked lakes fed by numerous streams carrying various materials. Then came a deposit of sandstone that covered the Entrada. This hard cap, the Curtis, provided the protection which retarded erosion and created the spectacular fluted walls that display the outstanding scenery of Cathedral Valley.*

JEFF GNASS

JOHN P. GEORGE

Colorful Bentonite Hills are *part of the Morrison Formation. The Morrison was laid down as volcanic ash, and is a formation in which dinosaur bones, sharks teeth, and many other fossils can be found. Although Capitol Reef is not as rich in fossils as some places in the West, this colorful formation can be recognized in many parts of the Colorado Plateau.*

ERCOSION is the LAST chapter in this story.

The Changing Picture

From this tranquil setting of swamp deposition, the picture changed gradually with the slow encroachment of the sea. This occurred in Cretaceous times, about 80 million years ago. It was during this episode that the drab Dakota Sandstone was laid down. This formation is composed of sandstone, conglomerate, and, in places, fossil shells, of which Oyster Shell Reef consists. The last two formations-the Mancos Shale and the Mesa Verde Sandstone—were deposited when the sea alternately advanced and retreated. The deposition of the Mesa Verde Sandstone ended this period, and would have nearly ended the geological story if it were not for one very important event that occurred about 60 million years ago.

At that time, constantly shifting plates of the earth's inner crust caused drastic landform changes. As the Rocky Mountains to the east were uplifted, the earth's crust wrinkled into a huge stair-step fold—the Waterpocket Fold. In conjunction with

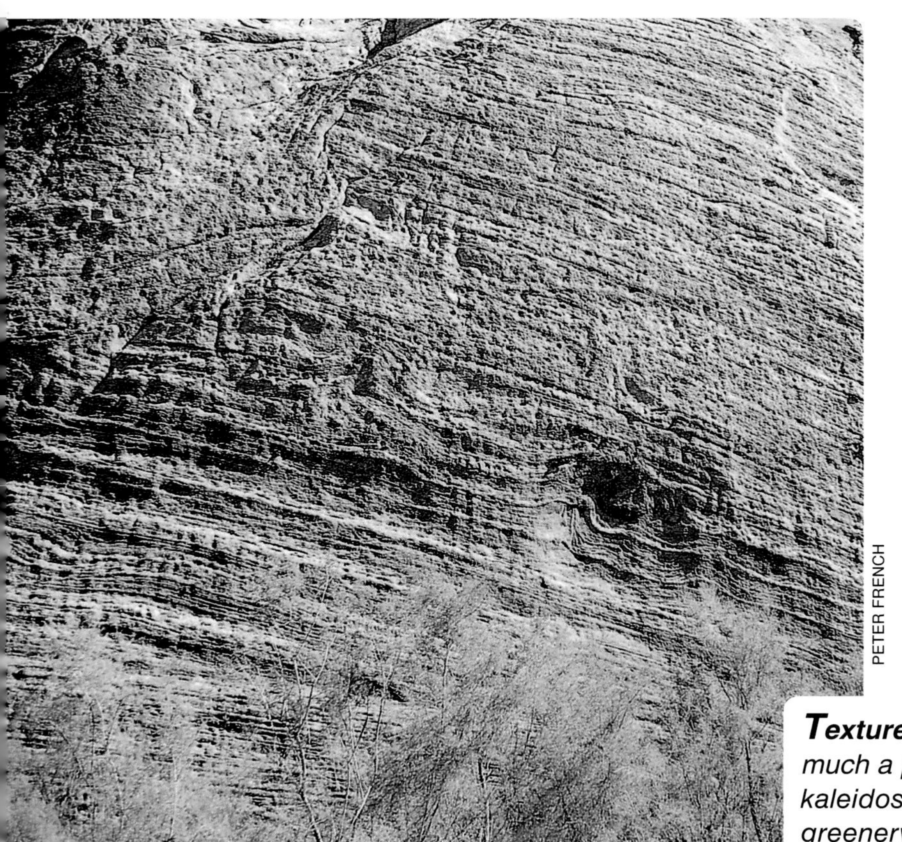

PETER FRENCH

Textures—in contrast or harmony—are as *much a part of the Capitol Reef experience as the kaleidoscope of colors and patterns. The gentle greenery of plants softens the harsh look of rock..*

The Summerville *Formation provided its own hard upper layer. Summerville tends to erode quickly but is usually readily recognized in cliff faces. It consists of siltstone , mudstone, shale, and sandstone laid down in an ancient environment of coastal tidal flats. Gypsum is fond in the Summerville which often has the appearance of chocolate-and a vanilla-banded ice cream cake.*

JOSEF MUENCH

The gray Mancos Shale reminds most visitors of a lunar landscape, and by moonlight it radiates its *coolest, most compelling beauty. Formed by deposition along the changing shoreline of a shallow sea, the formation is 3,200 feet thick at Bittercreek Divide. Its thick gray shale interfingers with thinner sandstone members like the Ferron and Emory, which contain seams of coal.*

GARY LADD

***Due to the uplift of the** Waterpocket Fold and the subsequent erosion of the Strike Valley (viewed here to the north), one can see each of the formations tipped up with just the edges exposed. In dry weather almost any vehicle can travel the road into the southern part of the park; however, if it is rainy even a four-wheel-drive can have trouble on the stretches of bentonite, which become very slick.*

this, the Colorado Plateau (of which Capitol Reef is part) was also gradually uplifted. Now exposed to erosion, the Colorado Plateau was sculpted into a colorful array of landforms. The Waterpocket Fold is distinctive among these—an eastward-tilted bend in the earth's crust that runs from Thousand Lake Mountain nearly 100 miles to the Colorado River. Capitol Reef contains nearly all of the Waterpocket Fold within its boundaries.

A Succession of Forces

To further complicate the geologic story of Capitol Reef, volcanic activity began again to the west about 20 million years ago. Tremendous lava flows welled up in the areas of Boulder and Thousand Lake mountains. These lava flows did not reach the Capitol Reef area, but lava boulders did, urged on by a succession of forces. Due to the extreme cold of one of the last glacial ages that covered parts of the North American continent, glaciers spread out over the lava flows of Boulder and Thousand Lake mountains. During the subsequent period of warmer climate, the melt-waters from the ice brought the lava boulders down into the area. These black boulders, rounded from many miles of rolling and tumbling, can be seen everywhere.

At about the same time that the lava flows occurred in the mountain regions to the west, molten rock was being intruded into cracks and fissures in the Cathedral Valley and South Desert areas. Here, dikes, sills, and volcanic plugs can still be seen, results of the erosion that has since taken place.

The uplifting of the land during the formation of the Waterpocket Fold was probably the point at which deposition ceased and full-scale erosion began. Because of the increased gradient of the few perennial streams and the large areas of now steeply-tilted land, erosion was stepped up to a degree which was never experienced during the time when the land was relatively level.

Now, the Fremont River, flowing from west to east, and Pleasant and Oak creeks, crossing to the south, are the only perennial streams found in Capitol Reef. Polk Creek, in South Desert, is a

Although there are several streams that cut through *the Waterpocket Fold, only Pleasant Creek, Oak Creek, and the Fremont River (pictured here) flow year-round. During seasonal changes the Fremont Canyon is a delightful place to observe the bright greens of summer become golds and browns in fall and, occasionally, white in winter.*

GLEN SHERRILL

perennial stream, but it sinks away into the sand after running for only a few miles. All surface waters are part of the Colorado River drainage.

Geologically speaking, erosion is the last chapter of this story—a story that is far from ended. Is not possible that one day in the remote, unforeseeable future, the sea may again cover the area and strange creatures may again tread the land? Standing in the midst of all this grandeur of eons past, which will surely change in eons yet to come, man's own life span seems very minute!

LARRY VENSEL

Named "Hamburger Rocks" for obvious reasons, these *rounded remnants are all that is left of the approximately 500-foot-thick layer of the Carmel Formation which was deposited on top of the Navajo Sandstone.*

SUGGESTED READING

ANDERSON, GLENN L., CUTRI, CHRISTOPHER. *The 275 Million Year Story of Capitol Reef*. Brigham Young University, 2001. DVD.

COLLIER, MICHAEL. *The Geology of Capitol Reef National Park.* Torrey, Utah: Capitol Reef Natural History Association

FILLMORE, ROBERT. *The Geology of the Parks, Monuments, and Wildlands of Southern Utah.* Salt Lake City: University of Utah Press, 2000.

LADD, GARY. *Landforms—Heart of the Colorado Plateau: The Story Behind the Scenery.* Las Vegas, Nevada: KC Publications, 1995.

SPRINKEL, DOUGLAS A., CHIDSEY, THOMAS C, AND ANDERSON, PAUL B. *Geology of Utah's Parks and Monuments.* Salt Lake City: Utah Geological Association and Bryce Canyon Natural History Association, 2003.

STOKES, WILLIAM LEE. *Geology of Utah.* Salt Lake City: Utah Museum of Natural History, 1986.

Cathedral Valley

STEVE MULLIGAN

Not to discourage visitors from *traveling to Cathedral Valley, but rather a word of caution: Be prepared! This serene view of the Temple of the Moon (left) and Temple of the Sun (right) can change quickly if the clouds darken and heavy rain should fall—Cathedral Valley roads can become impassable.*

TOM TILL

With the exception of the *volcanic formations, most all the scenic walls, temples, and monuments in Cathedral Valley are the result of the hard layer of Curtis Sandstone retarding the erosion of the underlying, softer Entrada.*

Just a few short years ago Cathedral Valley was virtually unknown, except by a few prospectors and herders who wandered through. It is still remote—only primitive roads lead into it, and four-wheeldrive is almost mandatory. Cathedral Valley's strange erosional patterns—exposed volcanic dikes, sills, and plugs-give the area the look of another world. In the late 1960s this area was added to what was then Capitol Reef National Monument. The purpose of the addition, from Thousand Lake Mountain to the original northern boundary of the monument, was to enclose and protect more of the Waterpocket Fold.

Further evidence *of nature's many quirks. Why do you suppose she left this wall standing when all around is leveled.*

LARRY VENSEL

Soaring Entrada *Sandstone monoliths seemed like Gothic structures to early visitors, who called this area Cathedral Valley. Depending on the time of day, coloration can move from tan to pink.*

RUSS FINLEY

Like additions were made to the east and south of the original monument with the same purpose in mind. Capitol Reef National Park now encompasses nearly all of the Waterpocket Fold—from Thousand Lake Mountain nearly to the Colorado River approximately 100 miles. When all boundary adjustments were made about 200,000 acres had been added to the original 37,000. Visitors planning an excursion to the Cathedral Valley area should check at the visitor center for road conditions and information.

JEFF GNASS

Just when one thinks they have seen it all, *something like the Glass Mountain appears. During the many periods when the area was covered by shallow lakes, evaporation would leave layers of salts and other materials. This extraordinary jumble of crystalline gypsum (selenite) is always a subject of visitor interest and park ranger protection concern.*

JEFF GNASS

The spacious South Desert, set against the *backdrop of the Henry Mountains, is a premier panorama at Capitol Reef National Park.*

Overleaf: *Time spent in the Capital Gorge area gives one a feel for the dramatic forces of nature that formed Capitol Reef National Park. Photo by Dick Dietrich.*

Although summer is hot
and the weather unpredictable,
it is perhaps the best time
to view the wildlife.

Life in the Canyons

LARRY VENSEL

***Spring in Fruita is colorful with all the fruit** trees in bloom. The fruit may be enjoyed by visitors when it is ready to pick.*

The visitor to Capitol Reef has a choice of a wide variety of things to do and see. Although the largest number of people visit Capitol Reef during summer vacation, it is not necessarily the only time, or even the best time, of year to visit. With the exception of early June, the summer months tend to be very hot, and temperatures easily reach above 100°F.

Seasons' Finery

From July through September, afternoon thundershowers are common, and the weather can be very unpredictable. Consequently, the visitor must take precautions when hiking, sightseeing, and camping. Flash floods can be expected in any of the washes and canyons.

The spring months of April, May, and early June reward the early visitor with the beauty of awakening nature. This is the time when the cherry, plum, apricot, peach, pear, and apple trees of the historic orchards are in magnificent bloom. Fremont cottonwood, exotic tamarisk, single-leaf ash, and many other trees are breaking out of their winter's garb to become green again.

Spring is the time for many flowers in riotous colors; they are more numerous, of course, in times of greater rainfall. The moisture provided by the Fremont River assures the local area of many species of flowers such as the legumes (vetches, locos, and the beautiful blue lupine), and penstemons (firecracker, low, and Palmer) common to most western areas.

Even the more arid parts of Capitol Reef produce a certain number of desert wildflowers. Two of these moisture-conserving plants are the many species of cacti, and the Fremont barberry or "algerita," whose foliage is modified into small holly-like leaves which prevent loss of moisture.

Capitol Reef lies basically in the Upper Sonoran life zone; its higher elevations lie nearly in the Transition or "ponderosa" zone. The dominant trees

JEFF GNASS

***October blossoms of rabbit brush provide a** contrast with the dark background. Cathedral Valley, in the distance, is still a remote area.*

ALL WILDLIFE is important and Must Be Protected

NPS PHOTO BY JOEL MUIR

The mourning dove is but one of many *species of birds that are drawn to the lush habitat of the Fremont River valley. The orchards provide some birds with food; insects are abundant for others.*

here are the pinyon and junipers which make up the "Pygmy Forest." Although one thinks of ponderosa and Douglas fir as existing only in higher country, they can be found in the upper canyons of Capitol Reef. The shade, moisture, and cooler air of the canyons create an environment which ordinarily would be found only at higher elevations; this is why some of the plants of the higher zones can be found here.

The Wildlife

With the coming of spring, wildlife becomes active again. The abundant moisture provided by the Fremont River attracts a surprisingly large variety of birds. The leafing out of the trees lure back the many species of warblers, northern oriole, black-headed and evening grosbeaks, the shy catbird, and the even shyer yellow-breasted chat (which can be heard day and night, but can seldom be seen), and many more.

On the desert, mockingbird and sage thrasher are common, and swooping down from the high cliffs come the white-throated swift and many species of swallows. Golden eagles nest on the high Wingate ledges, and can be seen flying majestically in high sweeping circles hunting food for their offspring snuggled cozily out of harm's way in some wind-protected, secluded niche. Hundreds of mountain and western bluebirds migrate through the area on their way to nesting sites in the higher country to the west of Capitol Reef.

Although summer is hot and the weather unpredictable, it is perhaps the best time to view the wildlife. Newborn fawns and their mothers are readily seen in the orchard and surrounding hills. Stretched out on the rocks, yellow-bellied marmots can be seen sunning themselves after a long hibernation. Rock squirrels hurry across the roads, carrying food to their dens. The smaller chipmunks and antelope ground squirrels scamper here and there.

Reptiles, too, are most active during summer. Many kinds of lizards scurry about or just bask in the warm sun. Most common are the long western whiptails, the eastern fence lizard, and the side-blotched lizard. Water snakes are common along the river. Gopher snakes and racers busily hunt mice and other rodents; as predators, they partially assist in maintaining the balance of nature.

Two types of poisonous snakes are found here at Capitol Reef; these should be respectfully treated. Both are varieties of the prairie rattlesnake: the small faded midget rattlesnake, only 12 to 18 inches long, and the larger Great Basin rattlesnake, 20 to 24 inches long. Neither are large as snakes go, but both can be dangerous if teased or molested. Given half a chance, any of these snakes will avoid humans.

Within the boundaries of a national park or monument, *all* wildlife is important and must be

GAIL BANDINI

***B*irds are not the only** *wildlife that thrive in the Fruita area. Fairly large numbers of deer enjoy the bountiful grasses, alfalfa, and fallen fruit. During pioneer times the deer presented no problems to fruit farmers because they were hunted.*

LYNN CHANBERLAIN

A golden-mantled ground *squirrel keeps a close watch for foxes and hawks.*

MARGARET LITTLEJOHN

Rock squirrels are one of the *larger rodents to be found in the Capitol Reef area. They can be seen in almost any rocky area throughout the park.*

MICHAEL S. SALAMANCHA

Many species of lizards abound throughout the area. None, however, are *quite as flamboyant as the collared lizard. Notice the adaptive coloration that allows this reptile to blend in with the surrounding lava rocks.*

GAIL BANDINI

***G**oatsbeard is one of the many species*
of the sunflower family.

NPS PHOTO

***T**his rosy finch is not disturbed by the proximity*
of the visitor center.

GAIL BANDINI

***T**he roots of the sego lily were a*
food source of native Americans.

GAIL BANDINI

***P**rince's plume growing in an ideal*
habitat with plentiful moisture.

MARGARET LITTLEJOHN

***M**any grasses grow in the area. Indian ricegrass was*
used as a food by the seed-gathering Fremont Indians.

LYNN CHAMBERLAIN

***T**he ring-tailed cat is not a cat but a member of the raccoon family. This little animal is very shy and also nocturnal, which makes it difficult to observe. Keep a sharp look out at night for a possible sighting.*

***A**lthough beaver are not common, occasionally they can be found working along the Fremont River. Their main food source here is cottonwoods and willows.*

MARGARET LITTLEJOHN

protected. Certain animals are sometimes regarded as pests, too lowly to consider; but these, too, have their niches in the ecosystem. By destroying one species, the natural balance is upset, to the possible detriment of other species.

Many of the animals at Capitol Reef are diurnal—they can be seen by day; but perhaps more of the animals found here are nocturnal, traveling and hunting in the cooler evenings or nighttimes.

Bobcats hunt for the black-tailed jackrabbit. The occasional cougar, his presence known only by his pad marks in the damp earth of a streambed, or by the remains of some unfortunate creature not quite quick enough, prowls the canyons and ledges at night. If, by some slim chance, you should see this phantom of the desert, be grateful for only a privileged few ever see a mountain lion. Some people who have lived in "lion country" all their lives have not seen one.

Perhaps you will also be lucky enough to see a ring-tailed cat (not really a cat at all, but a member of the raccoon family). His eyes are so sensitive that it is virtually impossible for him to come out of his den in daylight. This little hunter prowls the trees, ledges, cracks, and crevices looking for small rodents and fruit.

Two species of fox live in the area—the gray fox is found in the vicinity of the Fremont River, and the little kit fox is a desert resident. Both are beneficial in keeping the rodent population in check.

Although quite rare in the headquarters area, the coyote is seen occasionally throughout the park.

The air above is shared at night by many species of bats and by the common nighthawks; they jointly help control the thousands of insects attracted to the area by the moisture of the river.

Seasonal Changes

Summer is the time when the orchards begin to bear fruit. These orchards date back to the time of the pioneers, and thus have historic significance. The orchards are now owned and maintained by the National Park Service as part of the historic scene.

LARRY VENSEL

***If a photographer were to wait** for a rainstorm to take a picture like this, it could mean a long, long wait in this arid climate. However, having a camera along when it happens has its rewards as muddy water cascades off the canyon walls in numerous places.*

If spring is the most desirable time of year to visit Capitol Reef, fall is not far behind. Autumn is the ideal time to really get out and see the area. Crowds have thinned out once school has begun in most communities, and the weather is usually fair and pleasant. Warm days, cool nights, and very little precipitation is the norm.

September can still be quite warm, but October and November are ideal for hiking, sightseeing, and camping. Back country roads have "healed up" after the summer floods, and, except for being a little dusty, they are very serviceable. The many miles of hiking trails take you to interesting places such as Hickman Natural Bridge, Cassidy Arch, Whiskey Spring, Golden Throne, and many others.

Although winter in Capitol Reef limits what the visitor can do, it is nevertheless a fascinating time to visit. Snow is not unusual, but it seldom stays long. Fairly cold nights may be experienced, but, again, daytime is normally quite comfortable.

If you are one of the fortunate ones to see Capitol Reef after a heavy snow, you will be rewarded by some of the most remarkable color contrasts anywhere on earth. Colors deepened by the moisture from the melting snows-chocolates of the Moenkopi; grays, yellows, lavenders, and greens of the Chinle; reds of the Wingate, all frosted by a cap of pure white snow and enhanced by a deep blue sky-should satisfy even the most aesthetic of nature lovers.

SUGGESTED READING

ABBEY, EDWARD. *Desert Solitaire, A Season in the Wilderness.* New York: McGraw-Hill, 1968.

ALDEN, PETER, AND FRIEDERICI, PETER. *National Audubon Society Field Guide to the Southwestern States.* New York: Chanticleer Press, Inc., 2006.

FLEISCHNER, THOMAS LOWE. *Singing Stone, A Natural History of the Escalante Canyons.* Salt Lake City: University of Utah Press, 1999.

WARRINER, GRAY. *Watermark.* Camera One, Seattle, 2007. DVD.

WILLIAMS, DAVID B. *A Naturalist's Guide to Canyon Country.* Guilford, CT: The Globe Pequot Press, 2000.

TOM TILL

***H**ickman natural Bridge was named* for Joseph S. Hickman, an advocate for the preservation of Capitol Reef in the 1920s. A self-guided trail to the bridge winds past scenic and geologic points of interest, as well as reminders of the presence of the Fremont culture.

AUDREY GIBSON

***B**icycling has become very popular* again in recent years. Although not permitted off road, the new high-tech bikes that are available today allow many visitors to experience the solitude and enjoy the scenery from the back roads of Capitol Reef.

These families provided enough children to justify the building of a schoolhouse; it still stands. The last class was held in the school in 1941. Frontier history is still recent at Capitol Reef.

Before There Was a Park

The mysterious Fremont Indians are the first inhabitants of Capitol Reef about which anything is known, however little that may be. Petroglyphs carved on the rock walls of the canyons by these prehistoric people are abundant. Pictographs (painted drawings) are not nearly so numerous. But all, if they could be correctly deciphered, would give clues to the events which constituted and shaped the daily lives of these Indians— events they thought important enough to record for their descendants to read.

Unlike Indians in the Mesa Verde area, who lived in cliff dwellings, the Indians of Capitol Reef made their homes in pit houses. These were constructed by digging a pit in the ground and then placing lava boulders in a ring around the pit. These boulders formed the foundation for the rock walls which, together with posts at each of the four corners of the house, supported the roof. After many years the structure would collapse into the pit, leaving only the foundation boulders, visible today.

Storage structures also exist, tucked away in high niches. These little houses have caused much confusion! It was first thought that these tiny storage cysts were dwellings; the people who lived in them, then, must have been very small. Hence, "Moqui," early pioneer name for "little people," was applied erroneously to the structures, and they are even now called "Moqui huts." Remains of a pit house and storage cyst can be viewed from the Hickman Bridge self-guiding nature trail.

JEFF GNASS

***The Fremont Indians built their brush-** covered homes over pits dug in the ground, but tried to protect their grain and other foodstuffs in these rodent-proof granaries.*

JOSEF MUENCH

***G**eologist C.E. Dutton looked down into the Waterpocket Fold and*
marveled at the strange beauty there, but he did not venture in. His words as he stood on Thousand lake Mountain in 1880:

More recent inhabitants of the area were the small bands of Southern Paiute Indians who were present here from the 1600s to the time of the first pioneers. The Southern Paiute spent the summers in the Fishlake vicinity west of Capitol Reef. There they lived on fish, deer, and small game, migrating in winter to the lower and warmer elevations of Capitol Reef and the Colorado River.

Desert bighorn sheep were important to the Southern Paiute, just as they had been to the Fremont. It is possible, however, that for these earlier people, bighorn was not just a source of food and skins. They are depicted in petroglyphs on the walls of Capitol Reef canyons more than any other animal; whether it had some religious or other symbolic meaning is not yet known.

"The colors are such as no pigments can portray. They are deep, rich and variegated; and so luminous are they, that the light seems to flow or shine out of the rock rather than to be reflected from it." C.E. Dutton

The Explorers

Indians were not the onl y people to be drawn to the area by the ample supply of water. It was, however, quite late in American history before others appeared on the scene. In fact, Capitol Reef was the last explored

Were they just the result of man's desire to create, or were *these petroglyphs some form of communication? Whatever the reason, many hours were spent in the laborious process of pecking out these outlines of people (perhaps deities) and the many forms of animals that lived in the area.*

MICHAEL S. SALAMACHA

territory in the continental United States—about the mid-1800s. Many explorers passed close by, but they did little more than view the area from afar.

In 1853, in search of a transcontinental railroad route, Captain John W. Gunnison travelled through central Utah along portions of the old Spanish Trail. In October of that year, Capt. Gunnison and several members of his expedition were killed by Pahvant Ute indians. Although Gunnison's party never came into the area now known as Capitol Reef National Park, his expedition set the stage for another famous explorer who did enter the current boundaries of the park.

In a privately funded expedition from 1853-54, John C. Fremont continued to search for a transcontinental railroad route and followed Gunnison's previous path through central Utah. Leaving the trail at the Green River, his party headed southwest along the San Rafael Swell. In early 1854, he crossed over Thousand Lake Mountain and ventured into what is now Capitol Reef National Park. His expedition daguerrotypist, Solomon Nunes Carvalho, produced a daguerreotype image during their visit that has been identified as a formation in Upper Cathedral Valley known locally as "Mom, Pop and Henry".

When Major John Wesley Powell was exploring the Colorado River he saw the edges of Capitol Reef. He also saw the Henry Mountains, which he called "The Unknown Mountains," since they did not appear on any existing maps. Powell crossed the Fremont River near its source, but did not explore it.

It was Major Powell who gave the river both its names, although he didn't know it at the time. While exploring the Old Spanish Trail and finding evidence of Colonel Fremont's earlier cache, Powell named it the Fremont River. The men of his Colorado River expedition also knew the river by its more colorful name—"Dirty Devil." The story is that Powell's men saw the mouth of a river they approached. Hungry for fresh trout, the men asked those in the lead boat if it were a trout stream. The disgusted answer came back: "Naw, it's a dirty devil."

The river still goes by its two names, becoming

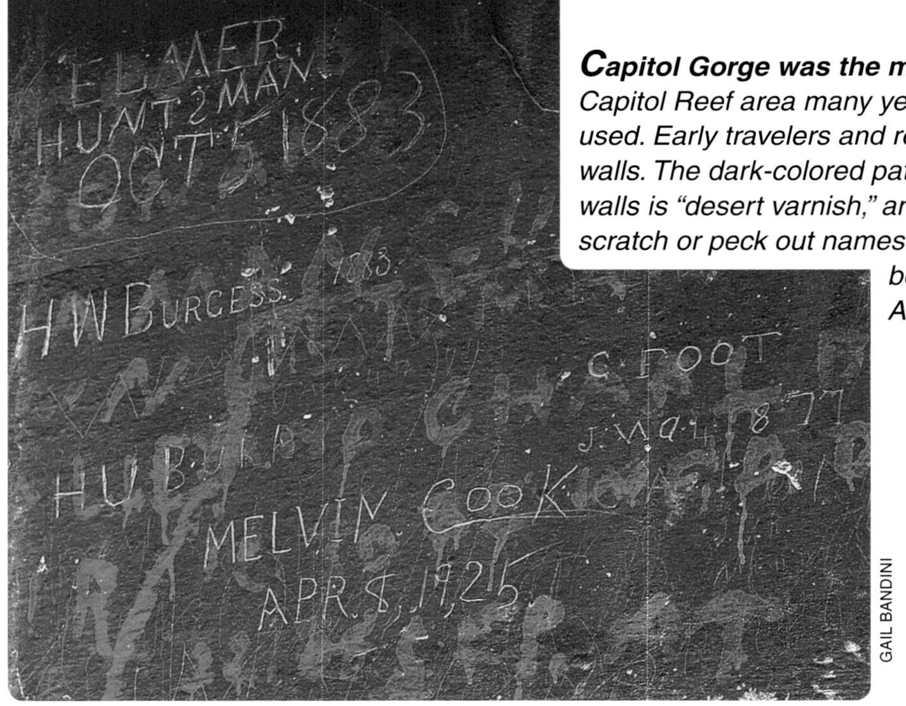

GAIL BANDINI

Capitol Gorge was the most popular route through the *Capitol Reef area many years before the Fremont Canyon was used. Early travelers and residents left their names on the canyon walls. The dark-colored patina or stain covering most of the canyon walls is "desert varnish," and made the gorge a favorite place to scratch or peck out names and symbols. Marks made today would be vandalism, but marks left by Native Americans and pioneers alike are history.*

indeed a dirty devil where Muddy Creek enters below Hanksville.

In 1872, some members of a Powell expedition, led by Almon Thompson, descended from Boulder Mountain along Pleasant Creek. They were the first scientific party to explore, map, and photograph lands that would become Capitol Reef National Park.

Many travelers skirted this area as they journeyed the Old Spanish Trail; but it looked dry and barren, and no one cared to attempt exploration of this forbidding terrain. Gilbert, a geologist, passed through the area on his way to map the Henry Mountains, but he had nothing of consequence to say about it. Later (1875), another geologist, E. E. Howell, made the first geological record of the area. His observations were the foundation for later studies.

In 1866, Captain James Andrus of St. George, Utah, organized a posse of sixty men to search for marauding Indians who had run off some cattle belonging to the Mormons. They found neither the Indians nor the cattle, but they did succeed in exploring some of the area. Having come over the top of Boulder Mountain and down into the Bicknell bottom, they followed the Fremont as far as the

Marks left by Native Americans and Pioneers alike ARE History

The rewards are always greater *for those who are willing to go a little further! This double bridge (perhaps one arch and one bridge), known as Brimhall Bridge, can be seen by hiking into the Hall's Creek area.*

TOM TILL

GARY LADD

***Many kinds of fruit** will grow in this oasis named Fruita by the nineteenth century Mormon pioneers. The Fremont River has deposited rich soil over the years and provides the water needed to irrigate the trees.*

Green River. Unfortunately, Captain Andrus' account has been lost, so we do not know what his impressions may have been.

Many travelers through Capitol Gorge carved their names on the wall in the narrows-a record now known as the "Pioneer Register." The earliest inscription was that left by J. A. Call and "Wal." Bateman when they passed through in 1871. These men are believed to have been prospectors looking for gold along the Colorado.

The Settlers

In 1873, Brigham Young, President of the Mormon Church, ordered the Richfield brethren to investigate the possibilities of grazing land in these valleys. They discovered good pasture and new farmland, so two years later the Richfield Cooperative Herd was moved into the area, wintering along Pleasant Creek and summering in the high valleys.

Soon settlers began moving into Grass Valley, the area surrounding Koosharem, and into Rabbit Valley along the upper Fremont River.

In 1880, Nels Johnson came to Capitol Reef and established his home near the junction of Fremont River and Sulphur Creek. This is now the picnic area of Capitol Reef National Park. He grew fruit there, and soon his home became a way station for weary cattlemen driving their cattle to summer and winter grazing lands. To these travelers he furnished board, lodging, and even wine made from the fruits of his vineyards. He also supplied prospectors who came during the San Juan gold rush of 1892.

Nels lived there until one day in 1902 when Sulphur Creek was in flood. A victim of an epileptic seizure, Nels fell into the swollen creek and drowned.

By 1881, Eph Hanks, a leader in the Mormon Church, had moved from Burrville and settled in a box canyon near Pleasant Creek. The surrounding cliffs enclosed his land, making a hothouse atmosphere ideal for growing fruit. Consequently, Hanks planted about two hundred fruit trees. Mrs. Hanks, viewing the lovely blossoms in the spring, named it "Floral Ranch."

Legend has it that Eph Hanks came to this area because he was a "polyg," as Mormons were then called. Hanksville, named for a cousin of Eph's, was established for the polygamists; the area here was too remote for the "feds" to bother with. If they did decide to try it, boys on fast horses could beat them, warn the Mormons, and all the "polygs" would be hiding out by the time the lawmen arrived. They hid

TOM TILL

MIKE HILL

***A**lthough small, all the necessities of an* *early school are here—from the well-used books to the handbell to call students to class. Built by Fruita residents in 1896, the school was opened officially in 1900. Early in its life it had only bare logs as interior walls; wainscoting and plaster were added in 1937. The school was in use until 1941.*

out in canyons such as "Co-hab Canyon." ("Co-hab" was another name then used for the polygamists.)

As more and more people moved into the area, the land became seriously overgrazed. One summer disastrous floods came, eroding the land so badly that several communities along the Fremont River were abandoned. Only Fruita, Caineville, and Hanksville on the lower Fremont were left inhabited.

At no time could Fruita support more than about ten families. These families provided enough children to justify the building of a schoolhouse; it still stands. The last class was held in the school in 1941. Frontier history is still recent at Capitol Reef.

Travel Hazards

Roads through Capitol Reef presented real difficulties. The first was just a trail down Sulphur Creek, and then down the Fremont River in a route similar to that of the present highway. This was a hazardous and time-consuming route; it crossed the river at least fifty times and cut through innumerable sand bogs.

MIKE HILL

***E**lijah Cutler Behunin* *hoped to tame the lower Fremont Valley for farming. He failed, but his house still stands.*

K.C. DENDOOVEN

***T**he Burr Trail Road* *winds precipitously to the crest of the Waterpocket Fold and leads to the Strike Valley Overlook and parking areas for those hiking Upper and Lower Muley Twistcanyons. Views of the Circle Cliffs are stunning.*

In 1884, Elijah Cutler Behunin, who built the first cabin in the area (still standing), decided to try to cross Capitol Gorge instead of negotiating the frustrating Fremont route. He struggled through with his wagon, moving large rocks out of the way. It took eight days to go the three and one-half miles. From then on all used this route, but not without hardship and danger.

Capitol Gorge can be completely dry, with no rain in sight; but rain on Miners Mountain can cause a flash-flood to come roaring through this gorge with little or no warning, rolling and grinding everything in its path.

Stand in the narrows and imagine a chocolate colored, eight-foot wall of water loaded with silt, boulders, rocks, sticks, and other debris rumbling toward you. Think with what apprehension the early pioneers must have traversed this road on a stormy day.

In 1962, the state of Utah constructed the present paved road through the area. With its completion, a reasonably safe crossing through Capitol Reef was available for the first time in history.

The Burr Trail also has an interesting history. Because of cattle grazing in the region of Grand Gulch to the south and Bullfrog Creek to the east of Capitol Reef, access was needed to the summer ranges of Boulder Mountain on the west. In the early days this access was provided by the Burr Trail, a series of narrow switchbacks climbing up the steep, unstable east slope of the Waterpocket Fold.

During the uranium prospecting of the 1950s, it was widened enough for vehicle travel. Today it is possible for modern automobiles, to negotiate this road—except during rainy weather. However, the steep gradient and unstable topography continue to subject the road to occasional slides and rock falls.

Just south of the junction of the Burr Trail and the Waterpocket Fold road is a landmark known simply as "The Post." This was a crossroads of sorts,

A walk *through the narrows of Grand Walsh with its polished walls is evidence of what flood waters can do. A native of the area told how she and her father were traveling through another nearby canyon when they heard the alarming roar of an approaching flood. Fortunately they were able to get their car out of the narrows and onto a bank before the swirling wall of water with its load of mud, rocks, and boulders went sweeping by.*

JOHN P. GEORGE

and cowboys of the late 1800s driving trail herds through Capitol Reef would rest or perhaps stop overnight at this point. Their horses were tethered to a large tree. Eventually the tree died, leaving only the trunk. In time, this "post" gave the crossing its name. Today all remnants of the tree are gone; replacing it is a small wooden pole with a sign on it reading, "The Post."

At one time the site also contained a small spring and a cabin. A local joke concerns a couple of travelers who stopped here to eat and rest. One fired up the cook stove to bake some biscuits, but when he opened the oven door, out jumped a badly singed skunk. Needless to say, the biscuits were forgotten in the mad rush to vacate the cabin!

Prospecting and Mining

During the "uranium boom" of the 1950s, many roads were built to give prospectors access to isolated potential sites of uranium ore. Much damage was done to the Capitol Reef area at that time, but fortunately many of the roads were built in the Moenkopi Formation. This surface, by its nature, erodes rapidly; thus most of these unsightly roads have nearly disappeared.

One of the uranium mines can still be seen in Grand Wash. This mine was first filed on in 1904, but very little ore was ever removed from it. Uranium mining in this area did not prove practical, as the ore had to be hauled all the way to Grand Junction, Colorado, for processing, a distance of 200 miles.

GARY LADD

The strenuous trail *on past Hickman Bridge affords serious hikers many splendid views of the Fremont Canyon. The scene below shows Sulphur Creek (foreground) joining the Fremont River. To the far right is the campground.*

Hiking and solitude go hand in hand at Capitol Reef. In less than a half hour the hiker is rewarded *with quiet and a chance to reflect—perhaps to imagine the hardships the early settlers must have faced in this ruggedland, or to recall colorful folklore about the infamous outlaws who used the blind canyons as hideouts and a network of trails as their escape route.*

MICHAEL S. SALAMACHA

Finally, when the Atomic Energy Commission decided enough uranium ore had been stockpiled, the government price support was removed. This event called a halt to a business that was already unprofitable in the Capitol Reef area.

Two other historical sites at Capitol Reef are the lime kilns used to manufacture lime for cementing materials, fertilizer, and whitewash. One of these, up Sulphur Creek, is not easily reached, but the other is near the campground on the Scenic Drive. The operation of these kilns was a back-breaking, tiresome job, as were many of the pioneer occupations.

The kiln consists of a large, hollowed-out hill with an opening at the bottom and another at the top. Limestone was gathered in the vicinity and stacked inside the kiln, leaving only enough room for firewood.

Pinyon logs were then placed in the remaining space and lighted. This fire had to be maintained continuously for 72 hours. At the end of this period the bottom and top openings were sealed. It was left to cool for several days, after which the kiln was opened and the lime taken out in the form of powder, much like the cement of today.

Many of the accounts of the early history of the old settlement of Fruita have been lost, but interested visitors can ferret out bits and pieces of information if they are persistent.

Colorful Outlaws

Capitol Reef does have its folklore—most of it concerns the colorful outlaws who made this their hideaway country. Books, such as *The Outlaw Trail* by Charles Kelly, *The Wild Bunch* by Pearl Baker, and *Robber's Roost* by Zane Grey, together with the movie, Butch Cassidy and the Sundance Kid, have immortalized the outlaw who used the Capitol Reef area as a hiding place and escape route.

It was part of the network of trails which Butch and his "Wild Bunch" used as they moved from Brown's Hole to Robber's Roost and south into New Mexico. Besides providing the only route west, the outlaws knew intimately all the blind canyons and all the water holes. They could safely go where many of the lawmen could not.

GLENN SHERRILL

Some of these canyons are barely wide enough for a horse and *rider, but the calming coolness of a sheltered canyon is a refuge on a summer's day. Water is rare in the Capitol Reef backcountry.*

A short trail in Grand Wash leads to a dugout, now ruined, said to have belonged to the Wild Bunch. It seems likely, since there is no other plausible explanation for a cabin's existence in this dry, remote area overlooking the trail. The dugout has burned, but some logs still lie there, partially covered by the dirt roof.

As the legend goes, Butch Cassidy was a real "Robin Hood," and he avoided hurting and killing. Although ever at odds with them, he never condemned lawmen for simply doing their duty.

One story is told of him as a young man caught at horse thieving. He went docilely along with the two lawmen who had nabbed him. At noon they

Looking east from the crest of the Waterpocket Fold, across *the Strike Valley and to the Henry Mountains, this dry, desolate stretch of terrain is as rough and remote as can be found anywhere. Imagine a cowboy rounding up cattle, or an outlaw trying to get to Robber's Roost beyond the Henry Mountains. What a place to be without water and a horse!*

stopped to eat, and the two deputies set about preparing lunch. Butch, handcuffed, stood by, waiting quietly. One man went to the creek for water; the other started the fire. As he bent over it, Butch shoved him, grabbed his gun, and had the second man covered as he returned from the creek. Butch took his gun, too, unlocked the handcuffs that still bound him, and rode off with his stolen horses.

Of course, Butch also took the deputies' horses, leaving them to walk the thirty miles back to town. But, in a gesture typical of him, he reappeared, gave the hapless lawmen back their canteens, and rode off again. He knew what it was to be out on the desert without water!

These are a part of the legends of the past; only a few written records give proof that they ever happened. Historical records and archaeological remains are few, and the missing pieces must often be guessed at.

Establishing the Park

A Presidential Proclamation, on August 2, 1937, set aside 37,000 acres to become Capitol Reef National Monument. Two subsequent Presidential Proclamations (1958 and 1969) enlarged the monument to 254,241 acres to include more of the spectacular Waterpocket Fold, thus expanding the area by nearly seven times. On December 22, 1971, an Act of Congress changed the status of Capitol Reef from a monument to a park; however, it also decreased the acreage. Now Capitol Reef National Park has 241,904 acres.

SUGGESTED READING

Cooley, John. *Exploring the Colorado River, Firsthand Accounts by Powell and His Crew.* Mineola, New York: Dover Publications, 1988.

Crampton, C. Gregory. *Standing Up Country: The Canyon Lands of Utah and Arizona.* New York: Alfred A. Knopf, 1964.

Houk, Rose. *Dwellers of the Rainbow.* Torrey, Utah: Capitol Reef Natural History Association, 1988.

Madsen, David B. *Exploring the Fremont.* Salt Lake City: Utah Museum of Natural History, 1989.

Martineau, La Van. *The Rocks Begin to Speak.* Wickenburg, Arizona: KC Publications, 1987.

Sanders, Ronald D. *Rock Art Savvy.* Missoula, Montana: Mountain Press Publishing Company, 2005.

SUGGESTED DVD

Heart of the Canyon Country, DVD #DV-33, 61 minutes, Whittier, California: Finley-Holiday Films.

All About Capitol Reef National Park

Capitol Reef Natural History Association

The Capitol Reef Natural History Association (CRNHA) was founded in 1963 as a nonprofit cooperating association. The association's purpose is to support historical, cultural, scientific, interpretive and educational activities at Capitol Reef National Park.

CRNHA operates the sales area in the Visitor Center and manages the Gifford farmhouse offering museum tours, daily craft demonstrations as well as selling items that relate to the historic scene of Fruita, Utah in the 1930's.

The CRNHA funds such programs as Frontier School Days and is a significant donor to the educational outreach program, oral history interviews, interpretive programs, and scientific research projects. When you purchase from CRNHA, you are supporting programs in Capitol Reef National Park.

CONTACT US

By Mail
Capitol Reef National Park
HC 70 Box 15
Torrey, UT 84775-9602

By Phone
Visitor Information
435-425-3791

By Fax
435-425-3026

Website
www.nps.gov/care

ROCK SQUIRREL
MARGARET LITTLEJOHN

CAPITOL REEF *Junior Ranger Program*

Capitol Reef National Park is a fun and exciting place to explore and learn about.

Interview a ranger, map an ancient earthquake, or get your feet wet watching waterbugs! Explore the park (with an adult), and learn about who found Capitol Reef first, how it got its name, who lived there first, and who lives here now.

Everyone can get into the act with a Family Fun Pack. Take your pick of several activities and get the whole family involved.

Visit the new Ripple Rock Nature Center, 3⁄4 mile from the visitor center. Check at the visitor center for facility hours and scheduled activities. Complete your Junior Ranger booklet and report back to a Park Ranger—say the park pledge, and **wear your Junior Ranger pin with pride!**

CAPITOL REEF NATIONAL PARK

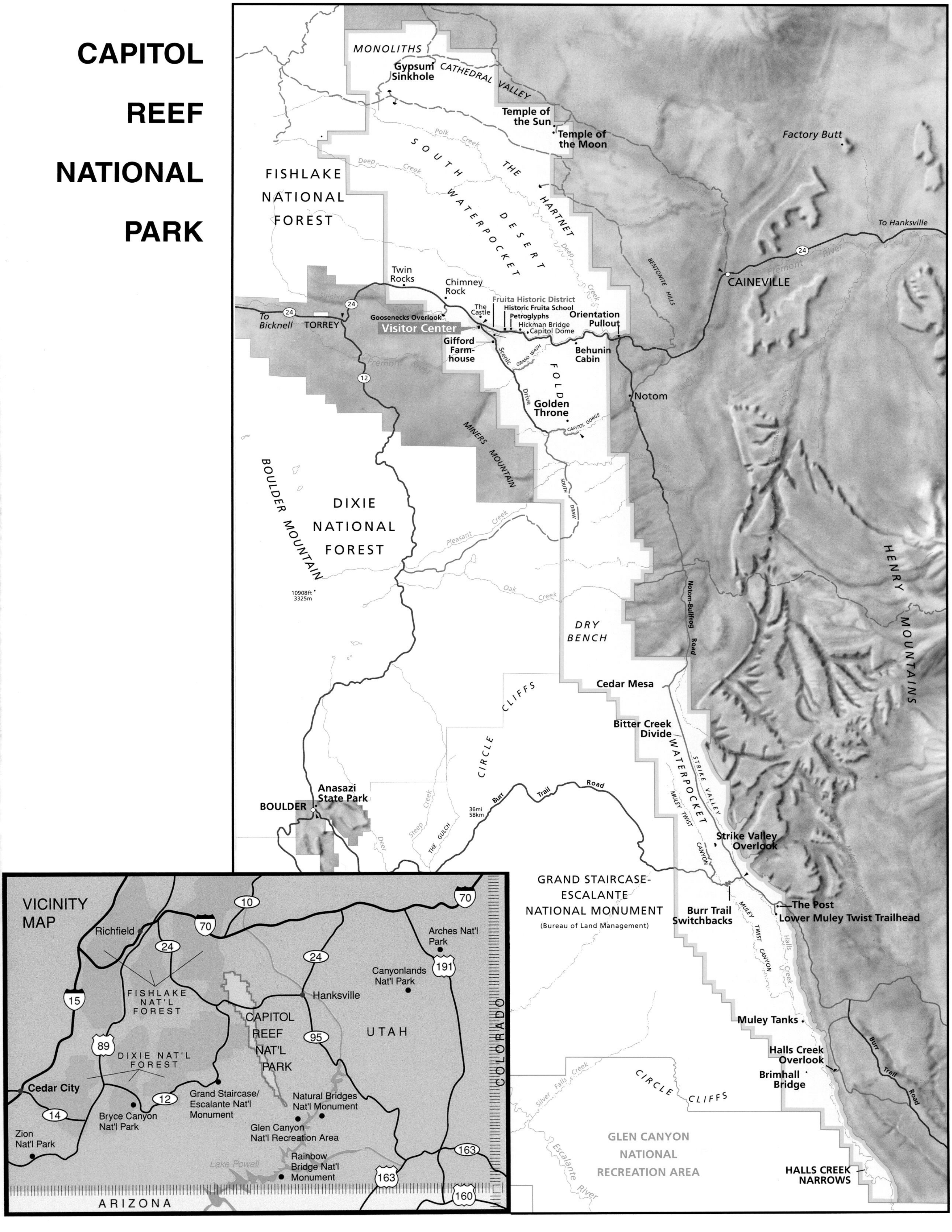

MONOLITHS
Gypsum Sinkhole
CATHEDRAL VALLEY
Temple of the Sun
Temple of the Moon
Factory Butt
Polk Creek
Deep Creek
SOUTH DESERT
THE HARTNET
FISHLAKE NATIONAL FOREST
WATERPOCKET
BENTONITE HILLS
To Hanksville
24
CAINEVILLE
Fremont River
Twin Rocks
Chimney Rock
Fruita Historic District
Historic Fruita School
Petroglyphs
Hickman Bridge
Capitol Dome
The Castle
Goosenecks Overlook
Orientation Pullout
Visitor Center
To Bicknell
TORREY
24
Gifford Farm-house
Behunin Cabin
Scenic Drive
GRAND WASH
FOLD
12
Notom
Golden Throne
CAPITOL GORGE
MINERS MOUNTAIN
BOULDER MOUNTAIN
SOUTH DRAW
DIXIE NATIONAL FOREST
Pleasant Creek
Oak Creek
10908ft 3325m
DRY BENCH
Notom-Bullfrog Road
HENRY MOUNTAINS
Cedar Mesa
CIRCLE CLIFFS
Bitter Creek Divide
STRIKE VALLEY
WATERPOCKET
MULEY TWIST CANYON
Anasazi State Park
BOULDER
Steep Creek
Deer Creek
THE GULCH
36mi 58km
Burr Trail Road
Strike Valley Overlook
GRAND STAIRCASE-ESCALANTE NATIONAL MONUMENT
(Bureau of Land Management)
Burr Trail Switchbacks
The Post
Lower Muley Twist Trailhead
Halls Creek
Muley Tanks
Halls Creek Overlook
Brimhall Bridge
Burr Trail Road
Silver Falls Creek
CIRCLE CLIFFS
GLEN CANYON NATIONAL RECREATION AREA
Escalante River
HALLS CREEK NARROWS

VICINITY MAP
Richfield
10
70
70
24
15
FISHLAKE NAT'L FOREST
24
Arches Nat'l Park
191
Canyonlands Nat'l Park
Hanksville
CAPITOL REEF NAT'L PARK
UTAH
95
89
DIXIE NAT'L FOREST
Cedar City
12
Grand Staircase/ Escalante Nat'l Monument
Natural Bridges Nat'l Monument
14
Bryce Canyon Nat'l Park
Glen Canyon Nat'l Recreation Area
Zion Nat'l Park
Lake Powell
Rainbow Bridge Nat'l Monument
163
163
160
COLORADO
ARIZONA

Capitol Reef Today and Tomorrow

Capitol Reef National Park presents us with a broad range of geologic happenings of millions of years ago, up to present day life in southern Utah. The Fremont Indians lived here, as did the Mormon settlers. Today people enjoy the picnic areas as a daily way of life, just outside the boundaries of the park.

Historic buildings are maintained, as well as the orchards that were planted by early pioneers. As we visit the scene—now—we must "see into the future." As a visitor, we have a wonderful chance to both enjoy what is here and take part in maintaining the way of life to be seen by generations to follow.

The adage—take nothing but photographs, leave nothing but footprints, applies significantly in dry desert climates. Each visitor should see Capitol Reef to the fullest, as well as seeing that nothing is disturbed in the process.

Let nature do its work of maintaining life in balance. Enjoy what you see, be sure the next visitor can enjoy the same.

FRANK BUCKLEY

This old barn is a reminder of Mormon pioneer days in Fruita.

KC Publications has been the leading publisher of colorful, interpretive books about National Park areas, public lands, Indian Culture, and related subjects for over 45 years. We have 5 active series – over 125 titles – with Translation Packages in up to 8 languages for over half the areas we cover. Write, call, or visit our web site for our full-color catalog.

Our series are:

The Story Behind the Scenery® – Compelling stories of over 65 National Park areas and similar Public Land areas. Some with Translation Packages.

in pictures... Nature's Continuing Story®– A companion, pictorially oriented, series on America's National Parks. All titles have Translation Packages.

For Young Adventurers® – Dedicated to young seekers and keepers of all things wild and sacred. Explore America's Heritage from A to Z.

Voyage of Discovery® – Exploration of the expansion of the western United States.

Indian Culture and the Southwest – All about Native Americans, past and present.

We publish over 125 titles – Books and other related specialty products. Our full-color catalog is available online or by contacting us: Call (800) 626-9673, Fax (928) 684-5189, Write to the address below, Or visit our web site at www.nationalparksbooks.com

Published by KC Publications • P.O. Box 3615 • Wickenburg, AZ 85358

Inside back cover: *"Sleeping Rainbow" was the Navajo name given the multi-hued landscape. Photo by David Muench.*

Back Cover: *Sandstone buttresses on the west side of the Waterpocket Fold. Photo by David Muench.*

Created, Designed, and Published in the U.S.A.
Printed by Tien Wah Press (Pte.) Ltd, Singapore
Pre-Press by United Graphic Pte. Ltd